YOUR KNOWLEDGE HAS VALUE

- We will publish your bachelor's and
  master's thesis, essays and papers

- Your own eBook and book -
  sold worldwide in all relevant shops

- Earn money with each sale

Upload your text at www.GRIN.com
and publish for free

Bibliographic information published by the German National Library:

The German National Library lists this publication in the National Bibliography; detailed bibliographic data are available on the Internet at http://dnb.dnb.de .

Imprint:

Copyright © 2019 GRIN Verlag
Print and binding: Books on Demand GmbH, Norderstedt Germany
ISBN: 9783668931794

This book at GRIN:

https://www.grin.com/document/458984

Marshall Goldberg

Are complexity and homochirality co-emergent phenomena?

GRIN Verlag

ARE COMPLEXITY AND HOMOCHIRALITY CO-EMERGENT PHENOMENA?

© 2019

M. Goldberg, MD

How did life begin? Why are amino acids left-handed (homochiral) in living organisms?

This paper suggests that these questions can be approached by analyzing computational constraints in complex dynamical systems (chemical or not) that are not in equilibrium, using the analogy of dynamic billiards on asymmetry graphs [1] based on Wolfram [2] cellular automata #30 and #110. It is proposed that complexity and homochirality are co-emergent phenomena based on the formulation:

$$\text{ASYMMETRY} + \text{ENTROPY} \rightarrow \text{COMPLEXITY} \,^{[1]}$$

where:

- 'Asymmetry' = any scale-free irregularity or non-homogeneity, e.g., pH, chemical concentration, temperature or any other parameter that has a gradient ($\nabla \Phi$).
- 'Entropy' = mixing, randomization or multiplicative noise.
- 'Complexity' = symmetry breaking, typified computationally by Wolfram cellular automaton #110 and dynamic billiards on its A-graph, #110.

Miller's [3] 1952 in vitro experiment demonstrated the abiotic production of amino acids in a putative early-Earth environment. Twenty amino acids in a 50:50, left and right-handed racemic mixture of stereoisomers (enantiomers), were produced by subjecting an oxygen-free mixture of water, methane, hydrogen, and ammonia to an electrical discharge. This was an important result but failed to explain why only Left-handed amino acids are associated with life. Miller's experiment resulted in racemic mixtures, indicating that the chemical reactions in his setup were symmetric, i.e. taking place in a system in thermodynamic equilibrium. Because Gibbs free energy, $\Delta \mathbf{G} = \Delta \mathbf{H} - \mathbf{T}\Delta \mathbf{S}$, (G—change in Gibbs free energy, H—change in enthalpy, T—temperature, S— change in entropy) is the same for both left and right enantiomers, only racemic mixtures are possible.

Frank's [4] amplification theory proposed that homochirality might have resulted from autocatalysis plus cross-chiral inhibition with an initial slight excess of one enantiomer (due to circularly polarized light, beta-decay, etc.). However, it has been suggested that natural racemization of these asymmetries occurs faster than the time required for the development of life, and the early prebiotic existence of both autocatalysis and cross-chiral inhibition is unlikely.

Several enantioselective chemical schemes have been developed to produce one stereoisomer for use as a bioactive product or drug. Phocomelia and other fetal abnormalities due to Thalidomide made it clear that one or the other enantiomer could be helpful or toxic. None of these clever methods has clarified the problem of the origin of life or why amino acids are left homochiral. During the past 50 years or so [5, 6], it has become apparent that symmetry breaking associated with the development of complexity depends on systems that are not in thermodynamic equilibrium—'far-from-equilibrium thermodynamically.' It has been shown that homochirality can be produced in a chemical system maintained far-from-equilibrium thermodynamically if there is a continual external energy input to the system and a continual decay of that energy. Generalized, abstract mathematical theories known as artificial chemistry have been helpful. This paper proposes an abstract geometrical theory known as <u>geometric chemistry</u>, based on dynamical billiards on asymmetry graphs derived from Wolfram cellular automata.

It has been suggested that [6] homochirality can occur in a 'noisy' thermodynamic system in which there is a:

- constant energy <u>input</u> into the system and;
- an equal energy <u>drain</u> from the system, so that;
- the system is maintained in a <u>far-from-equilibrium</u> state.

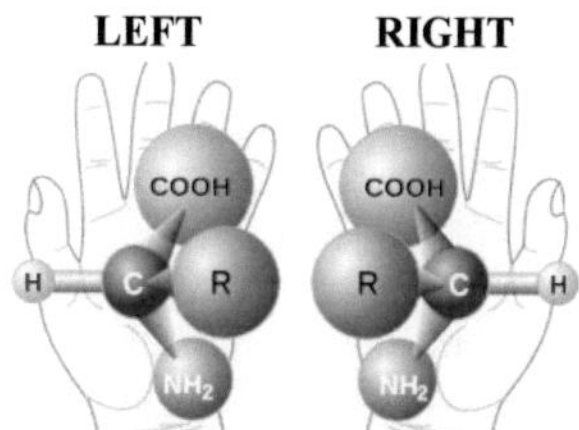

Figure 1 [4]

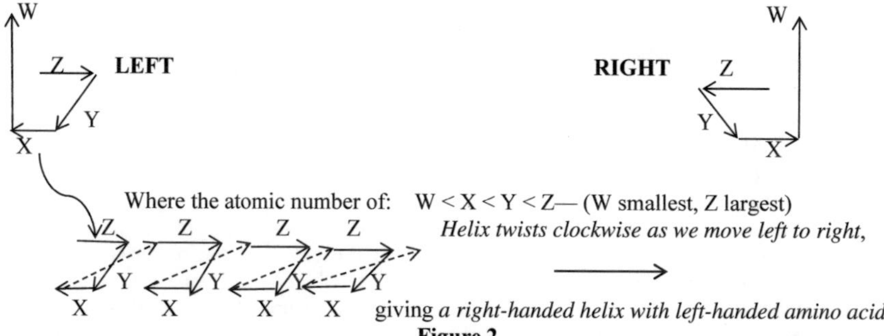

Figure 2

Figure 1 illustrates that the same amino acid has two enantiomers (stereoisomers), represented by left and right hands. The orientation of the molecules about the carbon center determines its chirality (handedness). **Figure 2** shows a simplified graphic form of Figure 1. Figure 2 relates the chirality to the order (clockwise or counterclockwise) of the arrangement of the atomic numbers of the molecules covalently bonded to the central carbon atom. Curl the fingers of the <u>left hand</u> with the thumb pointing straight up. Then the atomic numbers of the molecules added

3

to the apices of a carbon-centered tetrahedral molecule starting with the highest atomic number at the knuckle (metacarpophalangeal joint), then going clockwise as the atomic number of the molecules decreases towards the fingertip, with the lowest atomic number at the thumb. The right-hand orientation is defined by the same procedure using the curled fingers of the right hand. For amino acids, we have the set, S = [COOH, R, NH2, and H], with COOH (the carboxyl group) the highest atomic number and H (the hydrogen atom) as the lowest atomic number. The arrangement, from highest atomic number to lowest atomic number, defines a left or right carbon-centered tetrahedral molecule, following the Cahn-Ingold-Prelog system [7]. An alpha-helix with left-handed amino acids is a right-handed helix. The chirality of amino acids determines the chirality of the Alpha helix polymer (catalyst in autocatalytic set). Dashed lines represent peptide bonds. In Figure 2 note that Left-handed amino acids fit together to make a right-handed helix. The dashed line connects X=NH$_2$ of each amino acid to the Z=COOH of the next amino acid in a polymer chain. Reading left to right, following the arrows, traces out a right-handed polymer helix. A di-peptide covalent bond is formed between the carbon of COOH of one amino acid with NH$_2$ of another amino acid. A water molecule is removed to form the dipeptide bond.

Figure 3 shows a 2-dimensional top view of a carbon-centered (grey circle) tetrahedral molecule.

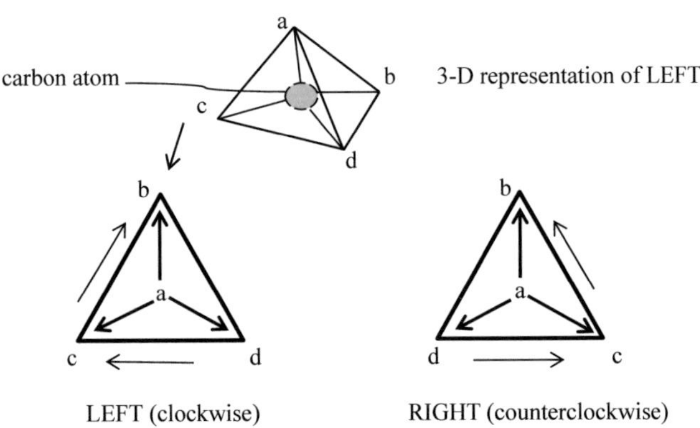

Figure 3

Figure 3 illustrates a 3-D representation of left, and also 2-D top views of left and right carbon-centered tetrahedral molecules. The carbon atom is in the center of the tetrahedron and is covalently bonded to other molecules at the apices of the tetrahedron. The apical molecule with the lowest atomic number is placed in the center of the 2-D figure. For a set of four atomic numbers, where d > c > b > a, the left side of the figure depicts a Left-handed molecule, and the right side of the figure depicts a Right-handed molecule. If two or more of the numbers in the set are the same, then the molecule is non-chiral or symmetric.

Figure 4 shows the asymmetry graph #110 (A-graph #110) derived from Wolfram cellular automaton #110, and the A-graph #30 derived from Wolfram cellular automaton #30. Recall from a previous paper [1] that cellular automaton #110 and its A-graph represent a complex dynamic system governed by the Principle of Computational Equivalence (PCE) and the Principle of Computational Irreducibility (PCI) [2]. Cellular automaton #30 and its A-graph characterize a random system. Asymmetry-graph #110 represents a complex small-world network, while A-graph #30 represents a random network. The two graphs are shown together to point out how they handle information flow differently. The dashed lines indicate that a random network can be transformed into a small-world network by increasing information flow at small and large distances. Increasing information flow between distant nodes is equivalent to having 'shortcuts' between those distant nodes. Decreasing information flow at intermediate distances is equivalent to reducing connections between nodes that are medium distances apart. These, are the very changes required in a random network to make it isomorphic with a small-world network [8].

COVALENT BOND FORMATION—REDOX POTENTIAL-**information flow**

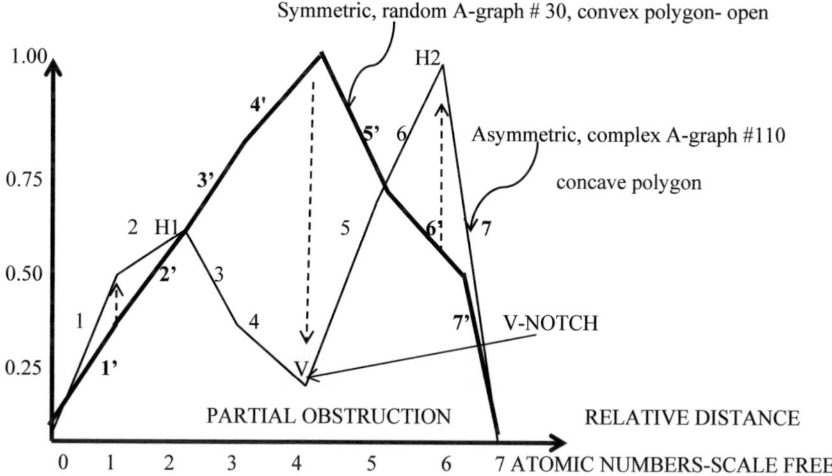

Figure 4

Figure 4 illustrates A-graph #110. It is a rational, <u>concave</u> polygon. Its internal angles are π/n, where n is a rational number. Note the V-notch which forms a <u>partial obstruction</u> to a dynamic billiard ball passing either way from H1 to H2 or from H2 to H1. A-graph #30 is also shown to illustrate the differences between the two graphs. <u>A-graph #30 is also a rational polygon, and mainly convex, thereby allowing, by contrast, free movement of a dynamic billiard ball from one side of the polygon to the other.</u> It is because of this difference in billiard ball freedom of movement that the random dynamics of A-graph #30 computes equal clockwise and counterclockwise (racemic) tetrahedral structures, while the computational consequence of the partial obstruction in the complex A-graph #110 means Left-handed tetrahedral structures are sequestered in H2, and separated from racemic structures in H1. <u>Thus, Left-homochirality is a computational consequence of complexity, and generally independent of any particular chemistry.</u> The numbers and primed numbers represent relative atomic numbers; the ordinate represents information flow and covalent bond formation. <u>The **computational** model is based on Figure 4:</u>

- Dynamic billiards [9, 10] on a billiard table (defined by random A-graph #30), models thermodynamic systems <u>at equilibrium,</u> and computes both left and right (<u>racemic</u>) carbon-centered tetrahedral molecules.
- Dynamic billiards on a billiard table (defined by complex A-graph #110), models a <u>far-from-equilibrium</u> thermodynamic system, and computes <u>Left homochirality.</u>

Referring to **Figure 5**, in dynamic billiards the billiard ball is assumed to be infinitely small, travel in straight lines without spin or 'English', and have its energy undiminished by collision with the 'wall' of the A-graph billiard table, i.e., the collisions are 'perfectly elastic.' meaning that the <u>energy of the ball is restored with each collision—a analogy with the idea of constant energy input to a thermodynamic system.</u> Moreover, the energy of the collision is used to supply the activation energy required to form a covalent bond at an apex of a carbon-centered tetrahedral molecule, i.e., there is a <u>constant removal or decay of energy.</u> The energy released (exergonic) from the billiard ball when it strikes a billiard table wall is coupled to the energy required (endergonic) for the formation of a covalent bond—a redox reaction. The angle of incidence of the ball with the wall is equal to the angle of reflection of the billiard ball from the wall, just as in real billiards. <u>By analogy with a chemical system, the computational system is maintained in a 'far-from-equilibrium' state.</u>

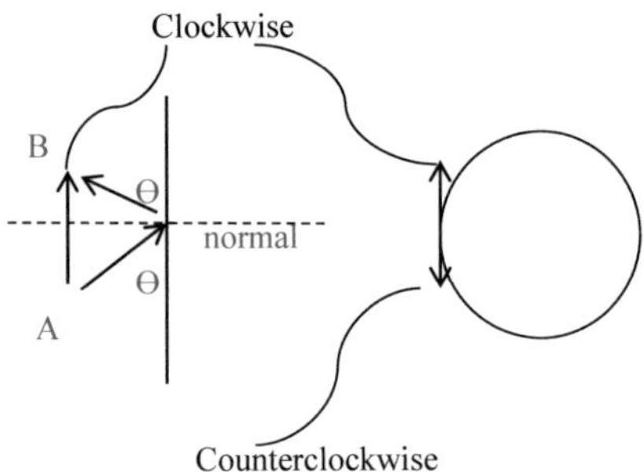

Figure 5

Figure 5 illustrates the 'mirror law'—the angle of incidence Θ = the angle of reflection Θ, and the direction of the path of the ball before it hits the wall relative to the direction of the path after it bounces off the wall is either clockwise or counterclockwise. The horizontal dashed line is the 90° 'normal' to the wall. For example, if the ball starts at 'A', strikes the wall and bounces off along path B, then the path rotates in a clockwise direction. If the ball were to start at 'B', strike the wall, and bounce off along path A, then the path rotates counterclockwise.

The relative clockwise or counterclockwise placement of covalent bonds at the apices of a carbon-centered tetrahedral molecule is determined by the respective rotation of the billiard path in the A-graph model. The path rotations are relative to the lowest atomic number which is placed in the center of the 2-dimensional representation of the tetrahedral molecule as seen in Figure 3. The walls of the billiard table are numbered. These numbers are taken to be the scale-free, _relative_ atomic numbers of molecules. When the billiard ball strikes any wall, the energy of the strike is used to form a covalent bond between a carbon-centered tetrahedral molecule and a _particular_ apex of the tetrahedron. The _particular_ apex is determined by the direction of the ball's path _before_ it hits the wall to the direction of the ball's path _after_ it hits the wall—i.e. clockwise or counterclockwise.

By convention, the smallest atomic number is placed in the center of Figure 3 above. Subsequent molecules that are covalently bonded to the apices of the carbon-centered tetrahedron are oriented either counterclockwise or clockwise with respect to the center of Figure 3.   For the present, a set of four different atomic numbers = [a, b, c, d] is used in accordance with the four covalent bonds of a carbon-centered tetrahedral molecule. However, the same system can be used to compute any asymmetric tetrahedral carbon molecule, not necessarily an amino acid. The same dynamic billiard model can be used to form larger molecules where, for example, we allow only sets with three numbers such as [a, b, c, carbon] where the fourth apical carbon bond is occupied by a second, carbon atom.   In this way we can model various polymers. In this paper our focus is on four-member sets (4-sets) for carbon-centered tetrahedral molecules.

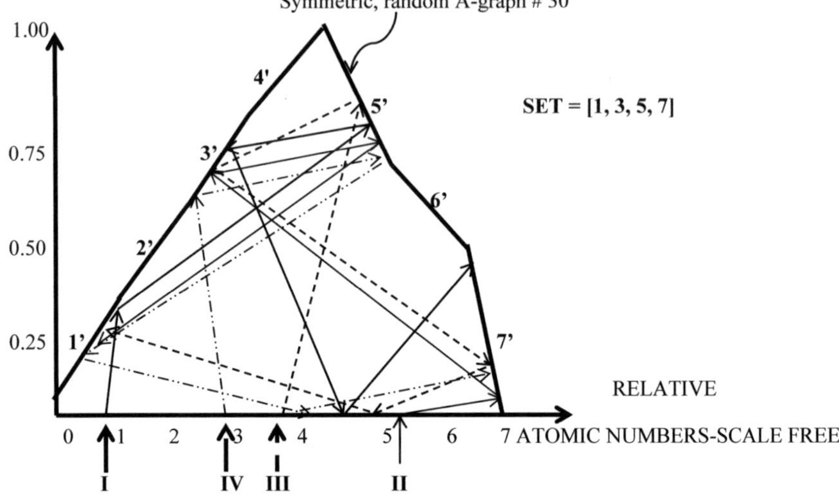

Figure 6

Figure 6 illustrates the computation of racemic (50:50) Left = Right enantiomers in A-graph #30. Pathways are adjusted for clarity. Roman numerals indicate the starting point of a billiard path.

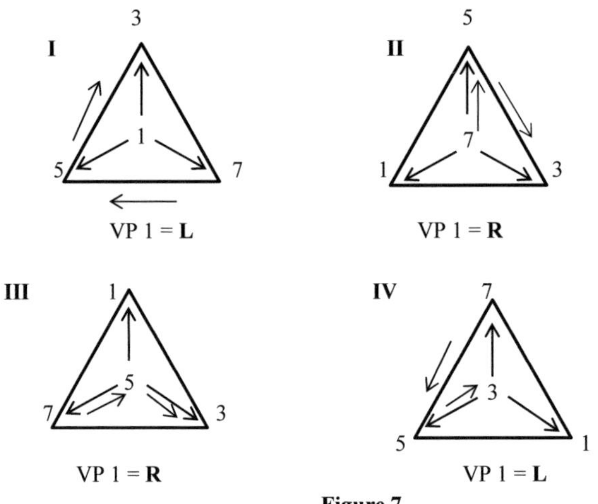

Figure 7

8

Figure 7 illustrates that the billiard paths in Figure 6 for the set [1, 3, 5, 7], compute equal amounts of left and right enantiomers. 'VP 1' means, 'from the viewpoint of #1, the lowest number in the set. Because A-graph #30 is 'open,' the billiard ball is more or less free to travel without obstruction from one side of the table to the other. In contrast, A-graph #110 is a concave polygon, such that the partial obstruction at the V-notch means that: **1.** the billiard ball cannot always travel directly from one side of the table to the other, and; **2.** if the billiard ball enters H2, then it becomes 'stuck' in H2, and; **3.** only Left enantiomers are 'stuck' or sequestered in H2.

COVALENT BOND FORMATION –REDOX POTENTIAL-INFORMATION FLOW

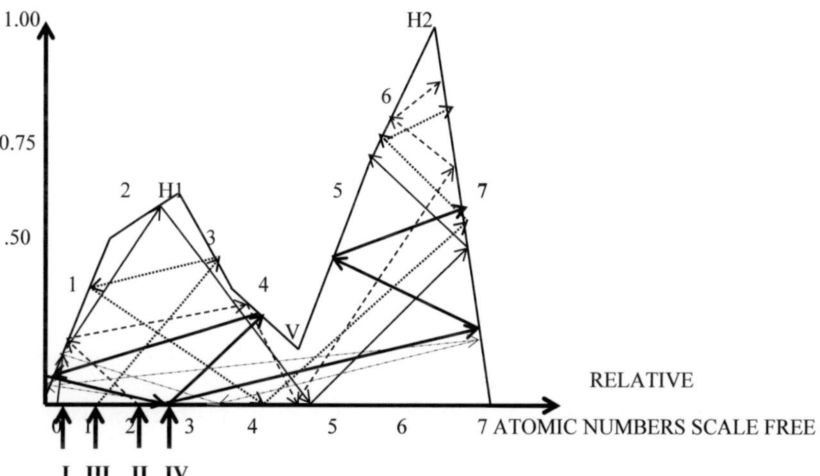

Figure 8

Figure 8 shows a dynamic billiard map for A-graph #110. The map shows only a portion of the possible map if one lets the map 'run' for a very long time. As the runtime increases, the density of the map increases. For the entire billiard phase portrait, P, we take a four-member subset S to define the chirality of an amino acid. Three representative areas are shown: **1.** in the H1 compartment, several lines are shown; **2.** the very fine 'dot-dot-dash' lines show representative pathways in which the billiard ball leaves H1, enters H2 and 'feeds back' to H1, avoiding the partial barrier created by the V-notch. The ball hits, e.g. wall #1 twice, computing a symmetric, non-chiral molecule. **3.** heavier solid, dashed, and dotted lines show billiard ball pathways which enter H2 and do not feed back to H2, but are sequestered in H2; these sets of the form, S = [a, b, (5 or 6), 7], become sequestered in H2 and represent Left Homochiral carbon-centered

9

tetrahedral molecules. Figure 8 illustrates that, given a set with two numbers in H1 and two numbers in H2, the billiard paths describe only Left homochiral molecules sequestered in H2. **<u>All such pathways give Left homochiral molecules</u>**. This is the case because, if we start in H1 with a lower number then go to a higher number in H1, we must rotate counterclockwise, then rotate counterclockwise to hit the abscissa, then rotate clockwise to hit '7' then rotate clockwise again to hit '5' or '6' giving us a Left homochiral molecule sequestered in H2. If we start in H1 with a larger number and then go to a smaller number in H1, we must rotate clockwise, then rotate clockwise from the smaller number to hit the abscissa, and then rotate clockwise again to hit '7', and finally rotate clockwise again to hit '5' or '6'. In both cases we obtain Left homochiral molecules sequestered in H2.

<u>There are two other cases</u>:

- With one number in H1 and three in H2, (this defines a symmetric, non-chiral molecule where at least two apical atomic weights from H2 are the same. [a, 5 or 6, 7, 7] **Figure 8** and;
- With three numbers in H1 and one in H2 (this defines a Left homochiral carbon-centered tetrahedral molecule. [a, b, c, 7].

 With two numbers in H1 and two numbers in H2, [a, b, 5 or 6, 7], where a>b or b>a, Left homochiral molecules are sequestered in H2 describing a Left-enantiomer. The sets are different, and represent various amino acids or any Left-enantiomer, carbon-centered tetrahedral molecule.

<u>In Dynamic Billiards on the A-Graphs of Wolfram Cellular Automata</u>:

1. A-graph #30 computes a 50:50 racemic mixture of Left and Right enantiomers within its single compartment beneath its single hill.
2. A-graph #110 computes a 50:50 racemic mixture of Left and Right enantiomers in compartment H1.
3. In A-graph #110, if the billiard ball starts in H1 manages to get past the partial barrier, enters H2, and manages to 'feed-back' to H1, it hits wall #1 in H1 twice, meaning that it computes a symmetric, non-chiral molecule.
4. <u>A-graph #110 computes (sequesters) only Left homochiral enantiomers in compartment H2</u>.
5. Both A-graph #30 and A-graph #110 compute non-chiral, symmetric molecules.

As shown in Figure 2, the left-handed enantiomers of an amino acid 'fit' together such that the 3-dimensional shape of the peptide-bonded polymer is a right-handed helix. Thus, the chirality of amino acids determines the chirality of the α-helix. This means that the catalysts in autocatalytic sets are right-handed helices.

Autocatalytic sets [12, 13] are considered important early steps in the establishment of life. These sets mimic a non-RNA/DNA world. They are protein-based systems with polymer-catalyst 'cores' (the 'genotype') that grow, reproduce, and evolve, plus other so-called 'food' molecules that express the information in the cores (the 'phenotype'). If molecules of various lengths and molecular weights are formed randomly as in A-graph #30 dynamic billiards (using n-sets with n>4), then the probability that large polymers may act as a catalysts increases exponentially with the molecular weight of the polymer. If large polymers (P1, P2, →P$_n$) catalyze a set of reactions: [N$_1$, N$_2$→N$_x$], (the nodes in the network), then the 'time' and 'energy' required for 'connections' among these nodes is minimized. In this sense, there is a shorter path, 'time-wise' and 'energy-wise' among what might have been distant nodes and we have the temporal/energy equivalent of a spatial small-world network. Thus, random events on A-graph #30 produce an autocatalytic set, and it is in this sense that a random network becomes a complex, small-world network (see Figure 4 illustrating the conversion of A-graph #30 to A-graph #110) [1].

In previous papers [1, 8] Wolfram cellular automaton #30 and its A-graph, and Wolfram cellular automaton #110 and its A-graph were compared to random and small-world networks, respectively. In a small-world network, there are 'shortcuts' between elements of the network. These 'shortcuts' decrease the number of nodes one must travel to get from a given node to a 'distant' node. They are spatial shortcuts. In autocatalytic networks, a catalytic reaction among molecules is a 'shortcut in time and energy' required to get from one molecule (network node) or molecular reaction to another, suggesting that autocatalytic food networks and complexity are also co-emergent phenomena.

Thus, a set of molecules at thermodynamic equilibrium (a random network) becomes a complex, small-world network (in a temporal/energy sense) as complexity emerges. A random network of molecules can become a complex, small-world network, i.e., an autocatalytic set according to the rule: asymmetry + entropy → complexity [19, 20, 21]. Imagine a random set of molecules formed in dynamic billiards in A-graph #30 allowing n-sets (where n>4). As the molecular weight of these polymers increases, there is an exponential increase in the probability that some polymers will catalyze the formation of covalent bonds between other molecules of the random set. The result is that a complex small-world network emerges as random A-graph #30 transforms into complex A-graph #110. Inasmuch as nucleic acids and sugars may be produced abiotically, and must fit together with L-amino acids and their right-handed alpha helix, then it does not seem unlikely, that a dynamic billiards computational model described in this paper can 'set' the chirality of life. Can nucleic acid-based molecules that are formed abiotically also 'self organize' into autocatalytic sets that reproduce, grow, evolve, and have phenotypes that express information in the RNA cores? From a computational perspective, both homochiral protein-first and RNA-first worlds are possible. The current dilemma over a protein or nucleic-acid-first world may be pointless [15].

Thus, <u>Left-homochirality and autocatalytic sets are a consequence of computation rather than any specific chemistry.</u> Therefore, life, and consciousness could be universal phenomena, and might well be found throughout the universe, manifesting themselves with any local chemistry.

Because, the complexity of cellular automaton #110 and its A-graph are universal (PCE), left homochirality may also be universal. Could it be that symmetry breaking, complexity, homochirality, autocatalytic sets, and life are inextricably bound together, and mutually co-emergent?

<u>Moreover, is the underlying reality of our universe computation</u> [22] <u>rather than any specific chemical process?</u> Do computational constraints associated with the emergence of complexity define the most general process by which complexity, life, intelligence, consciousness, and evolution, emerge? If <u>Asymmetry + Entropy → Complexity,</u> then life and all its manifestations may be a common feature of our universe, and not dependent on just water, carbon, and temperatures in the 'Goldilocks zone.' It could take any form consistent with the general computational boundaries imposed by A-graph #110 on the behavior of dynamic billiards. Could life on Pluto [18] take the form of large mats floating on a sea of liquid methane? Would their thoughts, not dependent on cell membrane ionic fluxes, compute fast enough in that frozen world so that we might communicate?

<hr>

References:

1. Five Papers: Goldberg, M. **1.** Classification of Cellular Automata Using Asymmetry Graphs. GRIN v341659; **2.** Synaptogenesis Grin v355574. **3.** '*Complexity Arising from Entropy Acting on Asymmetric Substrates.*' Telicom XII.22 June/July 1998, pp 38-40. Journal of the International Society for Philosophical Enquiry, ISSN: 1087-6456. **4.** 'Complexity' *Telicom* XV.19, September 2002, pp 44-48. Journal of the International Society for Philosophical Enquiry, ISSN: 1087-6456. **5.** *Complexity and Randomness*' *Telicom* XVI.3- June/July 2003 pp 65-67. Journal of the International Society for Philosophical Enquiry, ISSN: 1087-6456.

2. Wolfram, S., *A New Kind of Science*©2002 Wolfram Media Inc. Wolfram, Stephen, Wolfram Media, Inc., May 14, 2002, ISBN 1-57955-008-8.

3. Miller-Urey experiment. Wikipedia—many other references cited.

4. Two papers: **1.** Homochirality (Frank, Charles model 1953). Chirality (Chemistry). Wikipedia—many references cited. **2.** Chirality (figure of left and right hands) Wikipedia.

5. Enantioselective Chemical Synthesis. Methods, Logic, and Practice. Book. 2010, Corey, E.J.; Kurti, L. Evans, D. A.; Helmchen, G.; Rüping, M. (2007). "Chiral Auxiliaries in Asymmetric Synthesis". In Christmann, M. Asymmetric Synthesis – The Essentials. Wiley-VCH Verlag GmbH & Co. pp. 3–9. ISBN 978-3-527-31399-0.

6. Three papers: **1.** Noise-induced symmetry breaking far from equilibrium and the emergence of biological homochirality. Farshid Jafarpour, F., Biancalani, T., Nigel D Goldenfeld, N.D. Carl R. Woese Institute for Genomic Biology. Center for Advanced Study. **2.** Prigogine, Ilya; Nicolis, G. (1977). Self-Organization in Non-Equilibrium Systems. Wiley.   ISBN 0-471-02401-**3.** Wikiquote Ilya Prigogine—more papers cited.

7. Cahn-Ingold-Prelog Priority Rules, Wikipedia—many other references.

8. Goldberg, M. Grin v355574. Transformation of random network to complex, small-world network.

9. Bunimovich, L. Dynamical Billiards. Scholarpedia, 2(8):1813, doi: 10, 4249/Scholarpedia.1813. Georgia Institute of Technology, Atlanta, GA. (2007)

10. Stability and ergodicity of moon billiards.
Article *in* Chaos (Woodbury, N.Y.) 25(8):083110 · August 2015 DOI: 10.1063/1.4928594
Source: PubMed.

11. Theoretical models for the emergence of biomolecular homochirality. Walker, S. 2010.

12. Biosystems Chasing the tail: The emergence of autocatalytic networks. Volume 152, February 2017, Pages 1-10. Hordijk, W., Steel, M. doi.org/10.1016/j.biosystems.2016.12.002

13. Kaufman, S. (1993). The Origins of Order: Self Organization and Selection in Evolution. Oxford University Press. ISBN 0-19-507951-5. Wikipedia. Self organization.

14. Watanabe, A. et al. *Fractal and Small-World Networks Formed by Self-Organized Critical Dynamics*. J. Phys. Soc. Japan. 84, 114003 (2015) [10 Pages] http://dx.doi.org/10.7566/JPSJ.84.114003.

15. Two references: **1.** Wikipedia, RNA world; **2.** The RNA–Protein World www.ncbi.nlm.nih.gov › *Journal List* › RNA › *v.19 (5); May 2013*

16. Two papers: **1.** Artificial Chemistries, Banzhaf, W., Yamamoto, L. MIT Press, 2015. **2.** Artificial chemistries—a review. Artificial Life, 7(3):225-275, 2001.
17. Three papers: **1.** Wikipedia. Peptide bond. **2.** Peptide bond, Encyclopedia of Genetics, 2001. **3.** Right and left-handed helices, Boston University.

18. Does Pluto Have The Ingredients For Life? Howell, E. 2017. Space.com.

19. Watanabe, A. et al. Fractal and Small-World Networks Formed by Self-Organized Critical Dynamics. J. Phys. Soc. Japan. 84, 114003 (2015) [10 Pages] http://dx.doi.org/10.7566/JPSJ.84.114003.

20. Cohen, R., Havlin, S. Scale-Free Networks Are Ultrasmall. Physical Review Letters Vol. 90 Number 5 (week ending Feb. 2003). Minerva Center And Department Of Physics, Bar-Ilan University, Ramat-Gan, Israel.

21. Andresen, C. A. Properties of Fracture Networks and Other Network Systems. 2008 Thesis for PhD, Norwegian University of Science and Technology, Dept. of Physics.

22. Two papers: **1.** Gerard 't Hooft, The Cellular Automaton Interpretation of Quantum Mechanics—arxiv.org/abs/1405.1548.pdf. **2.** Information, Physics, Quantum: The Search for links, John Archibald Wheeler ('it from bit').

YOUR KNOWLEDGE HAS VALUE

- We will publish your bachelor's and
  master's thesis, essays and papers

- Your own eBook and book -
  sold worldwide in all relevant shops

- Earn money with each sale

Upload your text at www.GRIN.com
and publish for free